HOW CLEAN IS OUR CITY?

A Guide for Measuring the Effectiveness of Solid Waste Collection Activities

LOUIS H. BLAIR
ALFRED I. SCHWARTZ

THE URBAN INSTITUTE
Washington, D.C.

The Urban Institute is a nonprofit research organization established in 1968 to study problems of the nation's urban communities. Independent and nonpartisan, the Institute responds to current needs for disinterested analyses and basic information and attempts to facilitate the application of this knowledge. As part of this effort, it cooperates with federal agencies, states, cities, associations of public officials, the academic community, and other sectors of the general public.

The Institute's research findings and a broad range of interpretive viewpoints are published as an educational service. Conclusions expressed in Institute papers are those of the authors and do not necessarily reflect the views of other staff members, officers, or directors of the Institute, or of organizations which provide funds toward support of Institute studies.

Library of Congress Catalog Card Number 72-91412

ISBN No. 87766-052-2

UI 109-201-4

Available from:

Publications Office
The Urban Institute
2100 M Street, N. W.
Washington, D. C. 20037

List price: $1.95

Printed in the United States of America

CONTENTS

CONTENTS (Continued)

EXHIBITS

ACKNOWLEDGMENTS

This report extends the findings of a project undertaken jointly by The Urban Institute and the District of Columbia Government with partial funding support from the U. S. Department of Housing and Urban Development. This project was led jointly by Louis H. Blair and Alfred I. Schwartz of The Urban Institute. Harry Hatry provided general guidance. It is one of a series of studies on developing methodologies for evaluating local government services being undertaken by the Institute's State and Local Government Research Program.

The authors acknowledge with gratitude the many persons who provided major assistance to make the D. C. development tests successful: Alease Vaughn, Urban Institute, kept our data gathering and collection functions organized; Dr. David Seidman, our contract monitor with the D. C. Office of Budget and Executive Management (OBEM), was initially interested in the project and encouraged the support of different agencies within the D. C. government; Eugene Melgaard, D. C. Department of Environmental Services (DES), obtained and coordinated essential support from various offices in DES: Vijay Alsii, DES, helped develop mapping and analysis techniques; Albert Mindlin, Nathan Levy, and Pat Priester, Statistical Services Group, OBEM, and Dr. Beverly Causey, U. S. Census Bureau, contributed to the statistical design and analysis during the first phase of the study; James Alexander, Director, William McKinney, Paul Freese and Robert Perry, DES, all demonstrated continual interest to ensure that the measurement system would be implemented; and Kenneth Webb and Dr. Robert Sadacca, The Urban Institute, provided technical assistance.

This work was funded in part by the U. S. Department of Housing and Urban Development and in part by the District of Columbia Government. The views expressed are those of the authors and do not necessarily represent the views of The Urban Institute, the Department of Housing and Urban Development, or the District of Columbia Government.

FOREWORD

The cleanliness measurement system developed jointly by The Urban Institute and the District of Columbia Government has been and will continue to be of great value to our Solid Waste Management Administration within the Department of Environmental Services. The department has established levels of cleanliness at which streets and alleys are to be maintained, and is reallocating its refuse collection and street cleaning resources to maintain these levels throughout the city.

The department has also centralized refuse collection and street cleaning under regional directors, so that each director has under his control the resources necessary to maintain the cleanliness of his region. These directors are now being held accountable for the cleanliness of their regions, which will be determined by the litter rating measure discussed in this report.

On a broader front, the District Government will need to develop a performance measurement system as part of its response to the federal Office of Management and Budget Circular A-44. We are using the Department of Environmental Services as our pilot agency, partly because of the availability of the cleanliness measure as a prototype of a way of quantifying what previously had been regarded as a purely qualitative indicator of performance.

I believe the development of such measurement technology is a most valuable contribution to the management of urban governments. I would hope that continued emphasis will be placed on the development of such measures in those other areas of urban government which have not yet had the benefit of such an effort.

Graham W. Watt
Deputy Mayor,
District of Columbia Government

President,
International City Management
Association

Washington, D. C.
September 1972

SUMMARY

This report presents a set of procedures, developed and tested over a twelve-month period in the District of Columbia, to help a city or urban county assess the effectiveness of solid waste collection in its jurisdiction. More specifically the procedures are intended to help:

- Assess periodically the overall effectiveness of its refuse collection and street and alley cleaning programs – including citizen satisfaction with collection services;

- Evaluate the effectiveness of specific new or old collection programs;

- Allocate collection resources among service areas or neighborhoods of the city;

- Analyze alternative ways to provide services and improve conditions;

- Assess the performance of public or private collection forces (and perhaps help establish employee performance incentive programs); and

- Regulate private collection contractors.

These procedures can be used to identify needs and justify requests for additional forces and equipment when necessary. They provide for the collection of data by service area or neighborhood concerning the effectiveness of solid waste collection activities. How clean are streets and alleys? How frequent are health and fire hazards and odors occurring from uncollected refuse? How satisfied are citizens? How have conditions changed in the past year? These types of data are necessary for local government personnel to make well-informed decisions. To date, local government managers have had to rely on quantitative measures such as tons of refuse collected, or numbers of homes and businesses served. Such measures are inadequate because they provide little direct information on cleanliness, hazards, and citizen satisfaction.

Primarily, the procedures consist of three things:

- Periodic (two to four times per year) field inspections in which trained inspectors rate the cleanliness of preselected streets and alleys, using a four-point litter rating scale. The scale is expressed in a set of photographs used as reference for the various ratings.

- An annual citizens' survey to measure citizen perception of conditions and satisfaction with collection and cleaning services.

- An analytical capability to interpret the data collected, make estimates of conditions by neighborhood and for the whole city, determine changes in conditions over time, and recommend appropriate programs to correct deficient conditions.

INSPECTION OPTIONS

The report presents two alternative levels of field inspection intensity, one considerably less expensive than the other.

At the low-intensity level, a pre-selected random sample of about 30 to 50 block-faces (and, if desired, 30 to 50 alleys) are inspected two to four times a year in each neighborhood or service area. (Approximately six to nine areas are envisioned as the number needing to be distinguished in most jurisdictions.) Average street and alley litter ratings are calculated for each area and for the entire city. Ratings should be compared between areas and in the same area for different time periods.

The costs of this level of inspection effort, depending primarily on the number of service areas sampled, are expected to vary from $9,000 for four areas to $24,000 for 14 (Exhibit 2). Frequently, local governments will be able to use existing manpower for inspection operations and data processing, reducing actual added outlays.

At the high-intensity inspection level, one side of virtually all streets and, if desired, all alleys are inspected two to four times a year in those service areas predominately, moderately, or heavily littered (defined here as areas with an average rating in excess of 2.25). A sampling of 30 to 50 streets and alleys as described in the low-intensity option would be conducted in the generally clean service areas.

This higher level would also provide data for various operational purposes. Maps shaded to show the moderately or heavily littered areas of the city can be used to direct collection forces, to determine the amount of manpower required to clean an area, and to measure very small changes in the size of littered areas. Similar maps showing the locations of abandoned automobiles, abandoned housing units, or buildings with exterior code violations can be generated if systematic observations for these items are made during the field inspections.

This high level of inspection effort should provide sufficient information for middle-level management and control of forces. The costs of the high-level inspection are expected to vary from $16,000 for four service areas to $52,000 for 14. Costs will increase if data for other departments, as mentioned above, are collected. As with the low-cost option, often much of these costs can be offset by reprogramming existing personnel.

MINIMUM-COST CITIZEN SURVEY

The proposed "minimum cost" citizen survey provides a reasonably accurate way to measure citizen perception of conditions and satisfaction with cleaning and collection services. Responses can be compared over time to detect changes and to help evaluate programs. Telephone interviews with 100 citizens from each service area and in-person interviews with 50 persons without telephones selected in a statistically random way from the entire city will keep costs relatively low while providing what the authors believe are sufficiently representative responses from each service area. The interview is very short, requiring only about three minutes once the person has been reached by phone. The cost of a survey of 900 residents by telephone and 50 in person is estimated at $3,000.

ESTABLISHING A MEASUREMENT SYSTEM

To establish a measurement system, the following key actions must be taken:

1. Determine the proposed uses of the system.

   If it is for program and policy decisions and for budget

preparations, the low-intensity inspection option involving a relatively light sampling inspection of streets and alleys in each neighborhood should be sufficient. If it is also for direct management control of operational forces, the high-intensity inspection option involving the inspection of all streets and alleys in littered service areas and a sampling in the others is required.

2. Decide whether to inspect alleys as well as streets and whether to inspect private property as well as public property.

   If severe accumulations of litter in alleys and on private property frequently create hazards, then the jurisdiction should inspect them as well as streets.

3. Choose the items of data that should be collected on the field inspection.

   Suggestions are contained in Chapter II.

4. Prepare a set of photographs to serve as visual reference standards for the litter rating scale. *

5. Assign an analyst to direct the measurement system and analyze the data, and delegate at least two persons to work part-time as field inspectors.

   The inspectors must be trained and then checked periodically to maintain rating quality.

An operational measurement system could be established within three to six months from the time an analyst is appointed to work full time to implement the system.

---

* Photos developed for the District of Columbia are available from The Urban Institute.

I

INTRODUCTION AND OVERVIEW

BACKGROUND

In 1970, The Urban Institute published a report[1] proposing that local governments collect data to measure the effects of solid wastes and collection activities on such things as appearance, public health and safety, citizen inconvenience, and general citizen satisfaction with collection services. (The proposed measures are summarized in Exhibit 1.) It recommended that information be collected at least annually and be reported not only for the community as a whole but for individual service areas or neighborhoods within the community. A general set of procedures and estimated costs for collecting and analyzing data were presented.

The Urban Institute was invited by the District of Columbia to field test and implement a refined version of such a measurement system, supported in part by funding from HUD. The District's Department of Environmental Services (DES), the agency responsible for solid waste collection, the District's Office of Budget and Executive Management (OBEM), and the Institute jointly undertook this 15-month task beginning July 1970. The District provided two and sometimes three full-time personnel for inspections and vehicles for them, plus several part-time personnel to help design and evaluate the test program. Subsequently the Institute helped the District use the system to evaluate the effectiveness of "Operation Clean Sweep" (a special one-time clean-up program described further in Chapter V).

The full-system operational costs with two (semi-annual) field inspections of both streets and alleys and an annual survey of 900 citizens are estimated at approximately $39,000 per year, out of $17 million annual D. C. budget for solid waste management. Most of the measurement costs could be borne by redirecting the activities of existing personnel rather than by hiring additional personnel. A report of this work was presented to the District of Columbia Government in September, 1971.[2]

---

1. Blair, Hatry and Don Vito, "Measuring the Effectiveness of Local Government Services: Solid Waste Collection," The Urban Institute, October 1970, available through National Technical Information Service (NTIS), Springfield, Va. 22151.

2. Blair and Schwartz, "Measuring the Effectiveness of D. C. Solid Waste Collection Activities." Four volumes, August-September, 1971, The Urban Institute.

Exhibit 1. BASIC MEASURES OF SOLID WASTE COLLECTION EFFECTIVENESS

Health and Safety Measures

1. Rate of Observed Health Hazards and Injuries to Waste Collectors.

2. Rate of Observed Fire Hazards and the Number of Fires Involving Uncollected Wastes.

Aesthetic Measures

3. Rate of Observed Unsightly Appearances.

4. Rate of Offensive Odors.

5. Number of Objectionable Noise Incidents.

Citizen Inconvenience and Property Loss Measures

6. Index of Inconvenience.

7. Index of Damages Occurring During Collection.

General Public Satisfaction Measures

8. Index of Citizen Dissatisfaction.

The formal measurement procedures proposed in this report are expected to be of most interest to cities and urban counties with populations of 100,000 persons or more. However, the basic concepts should be applicable to smaller jurisdictions as well.

The measurement system can be useful whether services are provided by the jurisdiction itself, by private contractors, by regional authorities, or by some combination of these groups. For those jurisdictions which do not operate their own street and alley cleaning or refuse collection services, it could provide concrete evidence for seeking changes in contract collection services or providing for other regulatory activities.

OVERVIEW

This report extends the findings of the development work just described in the District of Columbia for use by local governments throughout the country. In the following chapters we point out the options and recommend to management the decisions it needs to make to establish a measurement system (Chapter II); how to implement the field inspection procedures (Chapter III); how to undertake the telephone citizen survey (Chapter IV); discuss and illustrate the uses of the data (Chapter V); and point out miscellaneous considerations in implementing the system (Chapter VI).

This report presents a set of tested procedures which a city or urban county can use to help:

- Determine the cleanliness of its streets and alleys.
- Assess the effectiveness of existing refuse collection and street and alley cleaning programs.
- Evaluate alternative actions for improving sanitation conditions.
- Allocate collection forces throughout the city.
- Estimate citizen satisfaction with cleaning and collection services.
- Measure the performance of public or private collection forces and help regulate private collectors.

- Establish performance contracts with sanitation union or private contractors.

- Justify requests for additional forces and equipment when necessary.

- Evaluate effects of special collection programs.

The procedures consist primarily of periodic field inspections, an annual citizen attitude survey, and an analytic capability to interpret the data collected and make recommendations.

Eight basic measures of waste collection effectiveness, shown in Exhibit 1, had been identified as important or potentially important in the initial report. It became evident quite early in the work with the District of Columbia that, of these eight, the measures of most general concern dealt with unsightly appearance (No. 3) and citizen satisfaction (No. 8). Also of considerable concern are measures dealing with health and fire hazards (Nos. 1 and 2) from wastes and abandoned automobiles. But data on these, as well as on some of the other measures, it turns out, could be obtained through the same procedures as those developed for data on unsightly appearance and citizen satisfaction. The main measures of effectiveness used in D. C. were:

- Average litter rating of streets and alleys by service area or neighborhood.

- Percent of streets and alleys with ratings of 2.5 or more by service area or neighborhood.

- Percent of citizens surveyed by telephone who were satisfied with various aspects of cleanliness and with household refuse collection services by service area.

The procedures described provide local government management with information on the effectiveness of collection programs that is needed to help make many major program and policy decisions. City and county executives and legislators, heads of sanitation, health, and fire departments, and budget officers have not had adequate information for many decisions on solid waste collection programs. Typically, they have used measures such as tons of refuse collected, number of homes or businesses served, and number of special pick-up collections provided to measure performance. But these measures indicate little or nothing about such effects of collection activities on the citizens and the community as cleanliness, health and fire hazards from uncollected wastes, excessive collection noise, and odors from uncollected garbage.

Some local governments have depended upon the random observations of inspectors or upon volunteered complaints. These are crude indicators. Because of their haphazard method of collection and qualitative nature, they cannot be relied upon to be sufficiently representative of conditions or attitudes throughout the community. Nor are they sufficiently precise to measure changes that may occur from one period of time to another, from one neighborhood to another, or from the implementation of particular programs.

We have not attempted to set standards for the level of service that should be provided nor to determine minimum acceptable conditions. Each jurisdiction should determine its own goals, set its own standards. This system can and should be used to define goals or standards and to measure progress made toward their achievement.

This report does not discuss ways in which conditions can be improved, how streets and alleys can best be kept clean, how health and fire hazards from uncollected waste can be eliminated. Nor does it identify the causes for severely littered conditions or low citizen satisfaction. Those considerations are beyond the scope of this work. However, analysis of the data collected can give insights into why unfavorable conditions exist. The measurement system is intended to help identify the status of solid waste collection system effectiveness, to identify problem areas, and to help assess the effects various programs have in alleviating problems.

II

ACTIONS TO ESTABLISH A MEASUREMENT SYSTEM

There are several key actions that management must take to determine the type of measurement system it needs and to establish the system. These should be made on the basis of the intended use of the system and the resources available.

1. Determine the purpose of the system.

If the purpose is primarily to help make program policy and budgeting decisions, a low-intensity inspection effort involving rating only a sample of streets and alleys semi-annually should be sufficient. If it is also for direct management control of operational forces, for providing detailed maps to other departments, and for obtaining actual counts and locations of unsatisfactory conditions, a high-intensity inspection effort is required in which all streets and alleys in the moderately and heavily littered service areas and a sample of streets and alleys in the other areas are inspected semi-annually or perhaps more often. Rough cost estimates for these options, as a function of city size, are given in Exhibit 2. The costs depend greatly on the number of service districts inspected.

2. Decide whether to inspect alleys as well as streets and whether to inspect private property as well as public ways.

We recommend that both streets and alleys be inspected; alleys are generally more littered than streets and littered ones usually harbor health and fire hazards. If severe accumulations of litter on private property occur frequently, creating health and fire hazards and harboring rodents, we recommend that the jurisdiction inspect private property as well (with a view towards initiating clean-up programs or forcing clean-up on private property).

3. Choose the items of data that should be collected as part of the visual inspection procedure.

Exhibit 3 shows many of the items that can be collected and our judgments on the importance of collecting each. Some of these items, such as building code violations and abandoned autos, would

Exhibit 2. ESTIMATED FIRST-YEAR SET-UP, INSPECTION, SURVEY AND DATA ANALYSIS COSTS [a]

City Size	Cost Element	COSTS [b]			
		Low Intensity Option Two sample inspection operations in each service area		High Intensity Option Two sample inspection operations in half of service areas and two full inspections in other half	
		Inspect Streets Only	Inspect Streets and Alleys	Inspect Streets Only	Inspect Streets and Alleys
100,000-250,000 Population 4 Service Areas 1500 Blocks	Set-Up and Equipment Inspections Analysis	2,500 1,000 1,500	4,000 2,000 3,000	4,000 2,000 3,000	6,000 4,000 6,000
	Total	$5,000	$9,000	$9,000	$16,000

4000 Blocks	Analysis	2,000	4,000	4,500	9,000
	Total	$6,500	$12,000	$13,000	$25,000
600,000-1,000,000 Population	Set-Up and Equipment	4,000	7,000	7,000	12,000
10 Service Areas	Inspection	2,000	4,000	5,000	11,000
6000 Blocks	Analysis	3,000	6,000	7,000	14,000
	Total	$9,000	$17,000	$19,000	$37,000
Over 1,000,000 Population	Set-Up and Equipment	5,000	9,000	9,000	17,000
14 Service Areas	Inspection	3,000	6,000	7,500	15,000
9000 Blocks	Analysis	4,500	9,000	10,000	20,000
	Total	$12,500	$24,000	$26,500	$52,000

a. Succeeding year costs for inspection and for analysis would probably be the same, but there would not be any substantial set-up and equipment costs.

b. The cost of one citywide citizen attitude survey would add $2,000 to total costs for the first two population groups, $3,000 for the third group, and $4,000 for the highest population group.

typically be of interest only to other departments, in which case the other departments should share inspection expenses. The inspection costs shown in Exhibit 2 are based on collecting only what are termed "essential" and "highly desirable" items from streets and alleys. Collecting additional data obviously will slow down the inspection operation and raise costs.

4. Select a set of photographs to serve as visual reference standards and permanent rating guides for inspectors.

The sets of photographs already developed and tested in the District of Columbia could be used or a set could be developed in accordance with procedures discussed in the following chapter. It might be appropriate to have a fifth category for very heavily littered conditions. The basic set of photographs (see examples in Exhibits 4 and 5) should not be changed for several years so that conditions readily can be compared over several years.

5. Decide how often inspections should be conducted.

At a minimum they should be conducted semi-annually—preferably once in the winter after leaves have fallen and the leaf collection programs have been completed, and once in the summer when odors are most prevelant. More frequent inspections might be needed if the inspection information is used for management control of work forces or if special cleaning programs or one-time cleaning operations are to be evaluated.

6. Assign an analyst to direct the measurement system and analyze the data.

The analyst should be responsible for training inspectors, monitoring their performance through periodic replications of their ratings, analyzing data, evaluating programs, and recommending new programs or operations where appropriate.

7. Delegate at least two persons to be field inspectors.

The amount of manpower expended on the inspection operations depends on the size of the city and the type of inspection system implemented. In the District of Columbia, only two inspectors, each half-time for the entire year, were estimated to be required for quarterly inspections of streets and alleys in the high intensity inspection option. In all cases, we feel it is important to have at least two trained inspectors so that a backup capability exists.

8. Determine the boundaries for the service areas or neighborhoods for which conditions are to be measured and reported.

Inspection results should be reported by service area or neighborhood as well as from the entire city. This will show differential impacts of wastes and collection activities on various neighborhoods and the different needs for collection services throughout the community. The operational inspection costs are directly related to the number of service areas. The costs were based on an assumption of about 500 to 1000 blocks in a service area or neighborhood. Many local governments have already established service areas or have political wards which could constitute such inspection service areas.

Exhibit 3. DATA THAT CAN BE GATHERED AS PART OF THE VISUAL INSPECTION (WITH SUGGESTED IMPORTANCE OF EACH ELEMENT)

Key:

*** Essential
** Highly Desirable
* Of Frequent Interest
– Might Be of Interest

Street - Block Face (The area between the center line of the street, the property line, and the cross streets at either end of the block)

***	1. Average cleanliness litter rating of street, gutter and sidewalk
**	2. Health hazards
**	3. Fire hazards
**	4. Abandoned automobiles on public property
–	5. Broken curbing
–	6. Broken sidewalks
–	7. Clogged basins

Private Property Exposed to Street

*	8. Heavily littered occupied property
*	9. Heavily littered vacant property
**	10. Health or fire hazards on occupied property
**	11. Health or fire hazards on vacant property
**	12. Abandoned automobiles on private property
–	13. Exterior housing code violations
–	14. Abandoned housing units

Alleys

***	1. Average cleanliness litter rating
**	2. Health hazards
**	3. Fire hazards
**	4. Abandoned automobiles
*	5. Abandoned major appliances or discarded furniture

Exhibit 3 (Continued)

Private Property Exposed to Alley

*	6. Heavily littered back yards
**	7. Health or fire hazards
**	8. Abandoned automobiles
–	9. Exterior building code violations

Waste Storage Areas

**	1. Overflowing containers
**	2. Improperly stored refuse
**	3. Flies, insects or signs of rodents
**	4. Strong odors
*	5. Spilled or scattered refuse
*	6. Uncollected refuse

III

VISUAL INSPECTION PROCEDURE

DESCRIPTION OF THE PROCEDURE

The heart of the measurement system is a visual inspection procedure in which trained inspectors rate streets and alleys for cleanliness from accumulations of solid wastes, and other items when desired. The inspector gives a numerical rating to the litter condition on a street or alley.[1] The rating is the basis for measuring differences and changes over time and among neighborhoods. A set of photographs of scenes of streets and alleys, scaled to illustrate the range of litter conditions, is used as a standard for assuring consistent ratings of litter conditions by different inspectors, and in different parts of the community over different periods of time.[2]

Exhibits 4 and 5 show examples of the range of litter conditions for streets and for alleys in the District of Columbia and the applicable litter ratings. Cities with frequent occurences of more severely littered conditions might provide photos for a condition "5". Cities with special types of litter – such as tumbleweed or palm fronds – might use different sets of photos.

---

1. The litter rating scale is a rough measure of the area of ground covered by scattered litter. We used a scaled litter rating rather than the square feet of area covered for two reasons. First, it is far easier to assign a rating from 1 to 4 than to estimate the area covered with litter. Second, it is believed far easier for the decision-maker to visualize what, for example, a condition "3" looks like than to visualize what say 10 percent of the surface area covered resembles. Volume of litter was not used. It would be difficult to measure and we felt that the area covered by litter was a much more relevant measure than volume. Weight was not used because the density of solid wastes varies greatly depending on the type of waste.

2. A set of standard photographs was developed in the District of Columbia. Four hundred photographs were taken of scenes representative of the range of existing litter conditions on streets and alleys, and the variety of backgrounds, structures, and land uses in which the conditions occur in the District. These were then judged independently by 19 persons who separated the photos into four groups corresponding to four different levels of litter conditions, with "1" representing the cleanest condition and "4" the most littered condition. Photographs on which there was complete or nearly complete agreement concerning its condition rating among 19 persons were included in the set used as the visual inspection standard. A set of these photographs was provided to each inspector for a reference standard. A lithograph set is available from The Urban Institute. A procedure similar to the one just described could be used by any jurisdiction to develop its own inspection reference standard.

Exhibit 4. EXAMPLES OF STREET LITTER CONDITIONS

Condition 1. Clean

Condition 2. Moderately Clean

Exhibit 4. EXAMPLES OF STREET LITTER CONDITIONS (Continued)

Condition 3. Moderately Littered

Condition 4. Heavily Littered

Exhibit 5. EXAMPLES OF ALLEY LITTER CONDITIONS

Condition 1. Clean

Condition 2. Moderately Clean

Exhibit 5. EXAMPLES OF ALLEY LITTER CONDITIONS (Continued)

Condition 3. Moderately Littered

Condition 4. Heavily Littered

The District of Columbia experience showed that the following inspection procedure is feasible and yields ratings that are reproducible. Inspectors rate the average litter condition of streets and alleys on a scale of 1.0 to 4.0 using the photos as standards. Intermediate ratings of 1.5, 2.5, and 3.5 are used when the degree of litter falls between the conditions illustrated.

An inspector makes a rating by driving over a predetermined route past each block face or through each alley to be inspected. He observes the condition and notes the block or alley identification, the litter rating, and the presence of any other item of interest. The data from each inspection can be recorded on a form such as the one shown in Exhibit 6.[3] For the type of data collected in the District of Columbia it was faster and less expensive to have the inspectors dictate their findings and location into a tape recorder for later transcription, rather than to complete the form by hand, which required the inspector to stop his vehicle. However, some jurisdictions may prefer to use a written report.

Tables would be prepared with figures showing the average litter ratings in each area (see Exhibit 7) and the summaries of other data that are collected. When all the streets and the alleys are inspected in the high-intensity option, maps should be prepared showing the level of cleanliness on a block-by-block basis. Maps in Chapter V illustrate how the data can be presented to show street litter conditions and locations of abandoned automobiles. The tabulated and mapped data should be compared from one inspection period to the next to see what changes have taken place over time.

An inspector should be able to rate about 20 block-faces an hour or 7 to 10 alleys an hour for a low-intensity sampling inspection operation where he often has appreciable travel time between block-faces. He should be able to rate about 40 block-faces an hour or 15 alleys an hour when all streets and alleys are being inspected in the high-intensity option and there is no appreciable travel time between inspections.

---

3. The instructions provided the inspectors to complete the forms are contained in Appendix A.

We recommend that ratings of street cleanliness be restricted to the block face – the area between the center line of the street on one side and the property line on the other and bounded by the cross streets on either end of the block. In the D. C. experiment it was judged unsafe for the inspector to drive his vehicle and, at the same time, try to rate both sides of the street. An alley, however, can be safely inspected in its entirety.

We recommend that inspectors note the presence of health and fire hazards caused by accumulations of solid wastes. A health hazard may be defined as a substantial amount of unenclosed garbage that rats and other rodents could readily feed upon or be harbored in. A fire hazard may be defined as a pile of combustible solid waste within five feet of a building which, if ignited, could cause substantial damage to the building. We observed in alleys a strong correlation between high litter ratings and the presence of health and/or fire hazards.

INSPECTOR SELECTION, TRAINING, AND REPLICATION

The District of Columbia experience showed that certain persons with identifiable characteristics could perform the inspection function adequately after a three-day training period. The quality of the inspectors is critical to the success of the inspection system. After training, most persons are capable of making the visual judgments required. Less certain and thus requiring extra attention is whether an inspector will maintain the required performance quality over a long period of time.

Formal education requirements are not critical per se, other than the ability to read simple instructions. Candidates for inspectors should be conscientious individuals known to be able to perform with alertness repetitive and often tedious tasks over sustained periods (up to five hours a day, five days a week), to read maps well, and to be observant. Each inspector has to drive, read the road map, and report his judgments on litter conditions and other observations using a tape recorder.

The inspectors do not have to be sanitation department personnel. They may be volunteer citizens or persons from other departments. (Perhaps the ratings may have a higher degree of public credibility if they are not conducted by sanitation department personnel.)

Exhibit 6. FORM FOR DATA COLLECTION

STREET/ALLEY CODE		DES AREA NUMBER		SERVICE AREA AND MAP NUMBER				TRACT GROUP NUMBER			TRACT NUMBER					BLOCK NUMBER				TRAVERSE NUMBER			BLOCK FACE		ALLEY NUMBER				INSPECTION DATE								
																													MONTH			DAY			YEAR		
1	2	3	4	5	6	7	8	9	10	11	12	13	14	15	16	17	18	19	20	21	22	23	24	25	26	27	28	29	30	31	32	33	34	35	36	37	38

Street & Alley Inspection ADP Coding Sheet

IME ECTION GAN ENDED ract)			AVERAGE CONDITION				LEAVES CODE		AVERAGE CONDITION WITHOUT LEAVES				CONDITION OF PUBLIC WAY				CONDITION BEYOND PUBLIC WAY				SPECIFIC ITEMS OR CONDITIONS CODES															CARD NUMBER	
																					0	1	2	3	4	5	6	7	8	9	10	11	12	13			
43	44	45	46	47	48	49	50	51	52	53	54	55	56	57	58	59	60	61	62	63	64	65	66	67	68	69	70	71	72	73	74	75	76	77	78	79	80

ES-13 T
(8/71)

Exhibit 7. ILLUSTRATIVE COMPARISON OF STREET LITTER RATINGS

Service Area	Average Sample Litter Rating[a]		Percent of Streets in Sample Rating 2.5 or Worse	
	June 1971	Dec. 1971	June 1971	Dec. 1971
1	2.43	2.11	15	7
2	1.80	1.79	0	0
3	3.13	2.53	61	22
4	2.10	2.32	10	25
5	1.58	1.44	0	0
6	2.71	2.76	31	38
Total[b]	2.38	2.27	29	25

a. These average ratings would be obtained by inspecting a minimum of 30-50 block-faces in each service area for the low intensity option and an average of 1000-1500 block-faces in each service area for the high intensity option. Statistical confidence limits for the averages should also be made available.

b. Weighted averages based on number of blocks in the respective services areas.

An annual three-day training period should be satisfactory. Potential inspectors would be trained; previously trained inspectors would receive refresher training. The first day the trainees would be introduced to the purpose of the inspection operations, to the photo standards, and inspection reporting procedures. The second day field inspections would be conducted with all trainees and the instructor in one vehicle. The third day field inspections would be conducted with each trainee driving a vehicle accompanied by an instructor. After the three days of training the trainees would be considered inspectors and would perform on their own. A more detailed description of the training program is included in Appendix A.

Eighty to 100 ratings made by each inspector during the first two weeks of on-the-job inspection and about 10 percent of his ratings thereafter should be replicated by a judge, preferably the analyst in charge of the operation, to see that each inspector is consistently making accurate ratings. If the ratings are inaccurate, the inspector should be retrained, or if that fails, replaced.

ESTABLISHING BOUNDARIES

For comparison purposes, the city should be divided into inspection areas. These can be service areas or neighborhoods of particular interest. Inspections are then made to estimate conditions for each area. The boundaries can be ward lines, census tracts, or service district lines. The District of Columbia, containing approximately 6,000 blocks, had already been divided into nine service districts, each containing between 500 and 1200 blocks. Many communities have already identified service areas for a variety of purposes.

The boundaries of the areas should be overlayed on a large-scale map of the city. Each block should be examined to determine if it is appropriate for inspection. Some blocks will be eliminated, perhaps because they are within a region in which the city does not provide refuse collection or street cleaning services. Each block-face and each alley remaining should be numbered consecutively. The number is a unique identifier which is used in the random selection of block-faces and alleys required for certain aspects of the measurement system operation.

TWO FIELD INSPECTION OPTIONS

A basic decision is whether to a) inspect a sample of streets and alleys, or b) inspect all streets and alleys in any given area.

To save time and reduce costs, only a sample of blocks and alleys needs to be inspected when the data is to be used solely for program and policy planning purposes. When the sample is selected in a statistically sound manner, considerable confidence in its being representative of the city can be obtained. The resulting inspection data will be useful for such purposes as determining which parts of the city are dirtier than others, for measuring the change in cleanliness over time, and for evaluating the broad effects of cleaning programs. It can be used to help determine and justify budget requests for additional forces as well as to help allocate forces among the various areas.

Inspection of all streets and alleys in an area requires substantially more time and expense. However, it has the advantage that maps can be produced showing litter conditions on a block-by-block basis allowing comparisons between areas and over time. Actual counts of items of concern such as those identified in Exhibit 3 can be obtained. This information can be used to direct field forces and to evaluate individual work crew performances. We do not believe it is necessary to inspect all streets and alleys in the predominantly clean areas.

In the District of Columbia sample inspections were made in those service areas generally recognized as clean to establish base conditions for comparison with other areas and to detect any significant deterioration in conditions. In the moderately and heavily littered areas, all streets and alleys were inspected to provide location information for the direction of cleaning crews as well as to evaluate changes on a block-by-block basis.

The two options are described in more detail below.

SAMPLE INSPECTION OPTION

The accuracy requirements and the resources available determine the size of the sample. We recommend generally that a sample of 30 to 50 block faces and 30 to 50 alleys be inspected in each area. For that sample size, if the average litter ratings of the block-faces (or alleys) in two different samples[3] are compared and the absolute difference is 0.25 or more, one can conclude with 90 percent confidence that there is a statistically significant difference in litter conditions for the two areas and that the difference is not due to the chance selections of certain block-faces (or alleys).[4] Appendix B provides a procedure for determining the sample size using this criterion.

The specific block-faces and alleys to be inspected in each neighborhood should be selected in a statistically random fashion. We have already recommended that each block-face and each alley should be uniquely numbered in each inspection area. Then 30 block-faces and 30 alleys should be selected using a set of random numbers. The block-faces and alleys selected should be marked on a map and given to the inspector. He should then plot a route between inspection points that minimizes total travel time. After the block-faces and the alleys have been inspected, the variances of the ratings (a statistical measure of the dispersion of the individual ratings about the average rating) should be calculated to determine if enough inspections were made. The specific procedure is discussed in Appendix B.

It may be desirable to know if statistically significant changes are occurring in the percentage of streets or alleys with ratings in excess of 2.5 or of any other percentages. Then the sample must be enlarged considerably to measure small changes in these percentages. Between 30 and 50 block-face inspections in an area will not provide a high degree of precision. For example, if the percent of block-faces with ratings of 2.5 or worse is 10 percent for one sample, there is a tangible possibility (10 percent), that, had all block-faces been inspected, the value would have been lower than 3 percent or higher than 17 percent, rather than 10 percent. With about 120 to 200 block-face ratings – either from combining results from four inspection areas or by more intense sampling – the range decreases to about 7 percent and 14 percent respectively. A formula to calculate the number of inspections required to meet accuracy requirements is given in Appendix B.

---

3. The samples may be taken from different service areas during one inspection operation to compare differences between areas or the samples may be taken from the same area at different periods of time to compare differences over time.

4. The accuracy of the ratings based on sample data is determined by the sample size selected and the ability of the inspectors to rate conditions precisely. The D.C. experience showed that, using a four-point scale with intermediate one-half point values, inspectors with acceptable performances could, on the average, rate conditions to well within one-quarter of a point of their actual value as determined by the judge.

INSPECTION OF ALL STREETS AND ALLEYS

This option calls for the inspection of one side of all the streets and, if so determined, all the alleys, in moderately and heavily littered service areas and a light sampling as described above in the clean areas at least semi-annually. Depending upon what additional data are collected, high-intensity inspection can provide actual counts and locations of abandoned automobiles (see Exhibit 8), health and fire hazards on public/private property, broken pavements and clogged storm sewer basins, and even abandoned houses and exterior building violations.

Street inspection routes and alley inspection routes should be plotted on separate maps directing inspectors to make ratings of every public, traversable alley and of one side of essentially every street in the area. The routes should be simple and planned to minimize travel time; they will be used for all subsequent inspections. The estimated costs of this option are presented in Exhibit 2.

ALTERNATIVE INSPECTION PROCEDURES

Alternative inspection procedures can be developed. One obvious procedure would involve inspection of all streets and alleys once every year or two and sampling inspections for the other inspection periods. The sampling procedure would be based on an analysis of the data collected in the first inspection. A sampling inspection procedure could reduce inspection costs, based on the D. C. experience, by as much as 50 percent. There is a loss in data and it would not be possible to produce all of the maps that the full-inspection procedure allows.

A cluster sampling procedure can be developed where adjacent block-faces are selected, rather than random selection of each block-face. This reduces travel time and can reduce overall inspection costs. A city probably would need to engage a consultant with statistical test design experience to develop a proper cluster sampling procedure.

SCHEDULING OF INSPECTION

Statistical analysis of the D. C. inspection ratings showed that there was no substantial variation in litter ratings from one work day of the week to another or from one time of the day to another time in the same day.

ibit 8. ABANDONED MOTOR VEHICLES LOCATED DURING VISUAL INSPECTION OPERATION

Key: ● on streets

▲ on alleys

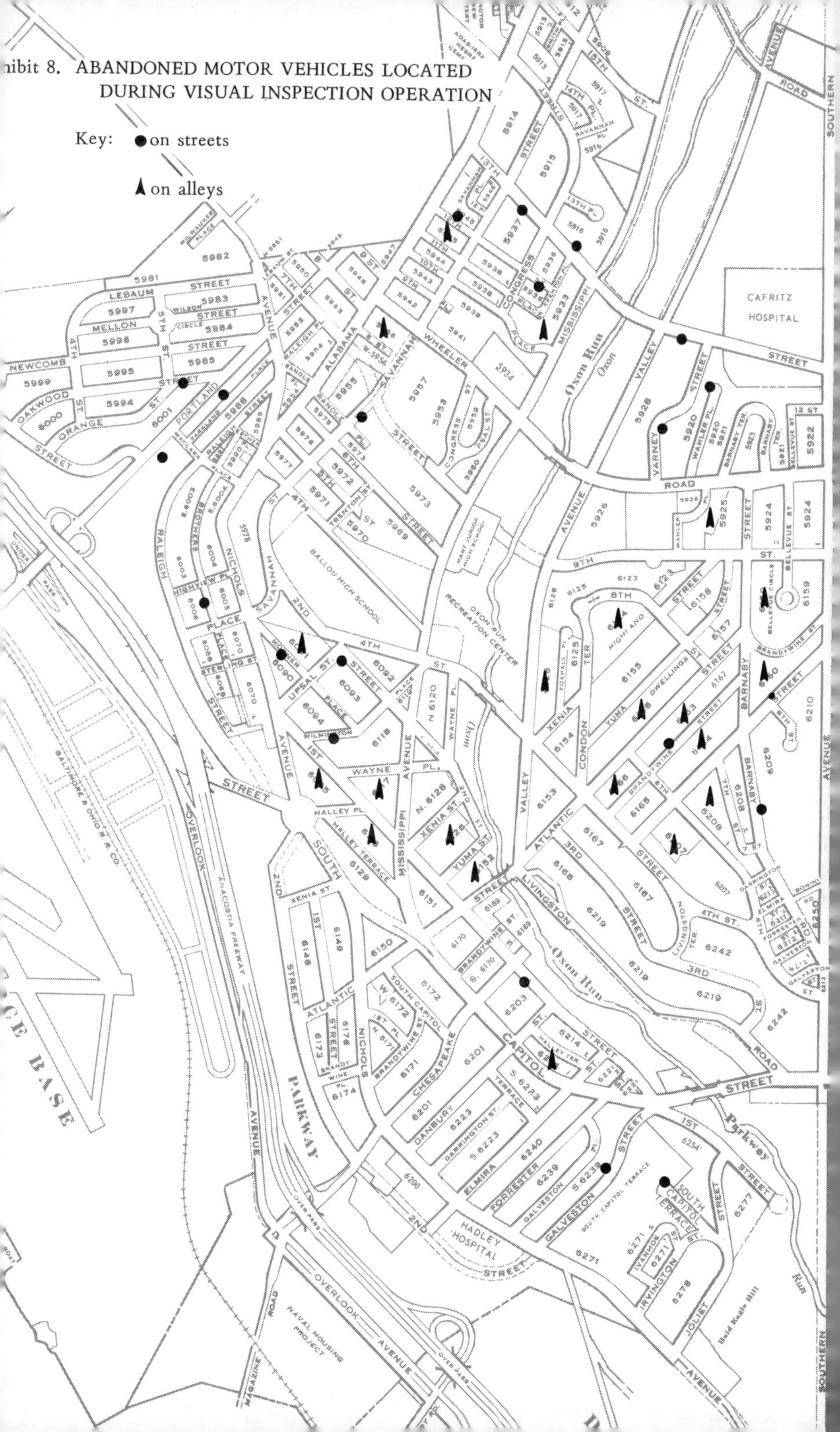

Inspections for street and alley cleanliness could be scheduled any day. We expect that the same stability in condition will be found in other communities and that this type of inspection can be conducted any or all days of the week. However, when the inspection system is used to evaluate the adequacy of collection frequency or whether or not collectors spill or scatter refuse, the timing of inspection must be related to cleaning operations.

IV

MEASUREMENT OF CITIZEN SATISFACTION WITH COLLECTION SERVICES

Public satisfaction with waste collection services is of major concern to local governments. Public perception of the quality of services provided can be partially measured by analyzing volunteered complaint data. However, complaint data is not likely to be representative of the whole community. A much better approach – though more expensive – is to measure citizen satisfaction by an annual citizen attitude survey. The authors recommend that both approaches be used.

ANALYSIS OF VOLUNTEERED COMPLAINTS

Volunteered complaints are rich sources of information on citizen satisfaction. All that has to be done is to record the nature of the complaint and the address at which each occurred and collate the data.

These data, however, have severe drawbacks. Dissatisfaction could well be more latent than expressed, so the volunteered complaints do not measure the full extent of dissatisfaction. Relatively few citizens bother to make formal complaints, and they may not be representative of a community's views. Some people do not complain because they have become conditioned to the presence of such things as odors and unsightly appearances. Complaints may be based on short-term or extreme incidents. Citizens might complain about specific items but still be generally satisfied with the services they receive. Finally, citizens may not complain because they do not know to whom to complain, do not believe it will do any good, or fear that they will suffer repercussions for having complained.

In spite of these drawbacks, complaints are useful as a quick indicator of change in conditions. When analyzed over time, they can show geographic areas of chronic complaints and perhaps poor service. In certain cases they can localize or point out particular routes where improvement is needed.

The authors suggest that volunteered complaint data be analyzed in the following way:

- A monthly tally of complaints by nature of complaint and by neighborhood or service area.

  This will give a rough indication of major problems within each area, and perhaps point out the need for additional services in some areas.

- Periodic map plots to see how complaints are geographically distributed.

  If they are localized, or repeated over time, an investigation can easily be made to see what remedial actions are required to improve conditions, if in fact investigation indicates that improvements are needed.

ANNUAL MINIMUM-COST CITIZEN ATTITUDE SURVEY

Why a survey? Because a survey measures citizen satisfaction and perception far more accurately than expressed complaint data. Government officials should not rely solely on survey data in making major decisions, but such data will provide information not otherwise available. It can be used to balance information from other sources. When compared with visual inspection ratings, it can help determine appropriate actions or trigger investigations into little recognized problems.

The recommended survey instrument is designed to measure, at low cost and with reasonable accuracy, citizen perceptions on:

- Cleanliness.

- Household refuse collection services, including missed collections, delays, spillage, and excessive noises by collection personnel.

- The prevalence of odors from uncollected solid wastes.

Exhibit 9 presents an illustrative comparison of responses to selected questions over a year's time.

Each survey involves telephone interviews of about 100 householders (or a mix of households and business establishments if desired) in each general neighborhood or service area plus in-person interviews of 50 persons selected from the entire city who do not have listed telephones.

The list from which a sample would be drawn for telephone interviews could be the semi-annual street address directory published by many telephone companies or one of the commercial street address directory companies such as Polk's or Hill's. Many householders are not listed in these publications because they do not have a telephone; have an unlisted, unpublished number; or have moved to their current address after data for the latest directory were collected.[1] The responses from these households may differ significantly from those from households with listed numbers. Consequently, in-person interviews using the same questionnaires with 50 persons from unlisted households throughout the city are recommended. The responses should be compared with the aggregated responses from the citywide telephone survey to determine if the findings from the telephone survey were representative of all residents, not just those with published numbers.

Primary emphasis is placed on a survey by telephone because it is considerably cheaper than an in-person survey of the same size, and it is likely to be considerably more accurate than a mail survey. With mailings, those <u>with</u> complaints are more likely to respond. If a large percentage of households in a community have no telephone, as many as 100 at-home interviews may be necessary.

---

1. There is growing interest in random dialing of telephone numbers for surveys which would involve residents with unlisted telephones. Such a procedure may prove better than the one described here.

Exhibit 9. ILLUSTRATIVE COMPARISON OF RESPONSES TO SELECTED QUESTIONS FROM ANNUAL CITIZEN SURVEY[a]

Question	Percentage Responses[b] 1971	1972	Statistically Significant Change
(1) What do you think of the cleanliness of the street you live on?			
a. Clean	14	13	
b. Mostly clean	45	50	
c. Fairly dirty	27	24	
d. Very dirty	14	13	
e. No opinion	0	0	
%Satisfied (a + b)	59 ± 8	63 ± 7	No
%Dissatisfied (c + d)	41 ∓ 8	37 ∓ 7	
(Number of respondents)	(100)	(110)	
(2) Have you noticed any change in the cleanliness of your streets in the last six months?			
Yes	43 ± 8	49 ± 9	No
No	52 ∓ 8	49 ∓ 9	
Don't know/no opinion	5	2	
(Number of respondents)	(100)	(110)	
(3) How often do the collectors spill trash and garbage or leave some that you have set out?			
a. Hardly ever	64	42	
b. Once a month	8	20	
c. More than once a month	23	36	
d. Don't know/no opinion	5	2	
%Satisfied (a)	67 ± 9	57 ± 8	Yes
%Dissatisfied (b + c)	33 ∓ 9	43 ∓ 8	
(Number of respondents)	(100)	(110)	

a. The same survey would be administered annually to different groups of citizens selected at random. Responses would be compared as shown to detect any significant changes occurring over time.

b. The ranges shown indicate the approximate statistical confidence limits (90%).

The suggested survey interview form is shown in Exhibit 10. It was developed from our experiences with the telephone survey undertaken for the District of Columbia. The D. C. experiment showed that a survey form similar to this one could be completed in about three minutes once the interviewer had the proper party on the line. However, there were difficulties in trying to reach some parties: telephones had been disconnected; householders had moved or could not be contacted; a few did not have time to participate at the moment called.

The respondents generally seemed interested and glad to participate and express their feelings. In view of the relative brevity of the survey and the willingness of respondents to talk, expansion of this survey to obtain information on other services such as public safety or health should be considered.

We recommend that an equal number of households be interviewed from each census tract to give a fairly uniform distribution of respondents by population. With 100 interviews per inspection or service area, the number of interviews per census tract is found by dividing 100 by the number of tracts in the area. Within each tract the specific block-faces for which respondents will be interviewed should be selected at random, (This can be done easily if all block-faces have been assigned unique numbers as was suggested for the visual inspection procedure; if the tract has a small number of blocks, a block-face is likely to be chosen more than once.) Three telephone numbers might be selected at random from the street address directory for each block-face selected, although only one response is needed from that block-face. The extra numbers are used if the first and second householders have moved, the phone has been disconnected, or a satisfactory response cannot be obtained. Even with only one or two numbers, a strong effort should be made to obtain a response from that block-face.

Initial calls to the subjects should probably be made between 5 p.m. and 9 p.m. on weekday evenings when most people seem likely to be at home. A strong effort should be made to interview the first person selected from the block-face even if it involves a number of callbacks to reach the person. If the first party cannot be interviewed (say after 5 attempts over a week's time) then the second party should be called. If no one can be interviewed, then three more householders should be selected and the process repeated until a householder can be successfully interviewed.

Exhibit 10: TELEPHONE SURVEY FORMAT

Name ______________________ Sex: M☐ F☐ Date ______________

Telephone# ______________ Time at start __________

Address ________________ Time at end __________

Block-Face# ______________ Interviewer __________

Census Tract# ______________

Service Area# ______________

If unsatisfactory response, why:

_______	No answer.	_______	Person no longer at this addre
_______	Telephone disconnected or temporarily out of service.	_______	Person called not in.
		_______	Could not communicate.
_______	Repeatedly busy.	_______	Person called could not answ at this time. Call back later.
_______	Person no longer at this number.		

"Hello, may I speak with Mr. (or Mrs. if woman answers telephone). I am working with the city of ______________. My name is ________. I am conducting a survey to see what citizens think of the street cleaning and household refuse collection services. Would you help me by answering a few short questions? Fine."

First, do you still live at (give address) ? (If yes, continue. If no, terminate.)

Now, I will ask you a question and read you several answers. Then I would like for you to tell me which answer you think is the best or most appropriate.

1. What do you think of the cleanliness of (Insert the name of the street the person lives on) Street? Do you think the street is:

_______ Clean

_______ Mostly clean, but dirty in a few places

_______ Fairly dirty

_______ Very dirty (_______ Don't know or no opinion)

2. What do you think of the cleanliness of the alleys in your neighborhood? Do you think the alleys are:

_______ Clean

_______ Mostly clean, but dirty in a few places

_______ Fairly dirty

_______ Very dirty (_______ Don't know or no opinion)

3. What do you think of the overall cleanliness of your neighborhood?

_______ Clean

_______ Mostly clean, but dirty in a few places

_______ Fairly dirty

_______ Very dirty (_______ Don't know or no opinion)

4. Have you noticed any change in the cleanliness of streets and alleys in your neighborhood in the last 3 or 4 months? (If response is "No", go to #6 or #8)

_______ Yes

_______ No (_______ Don't know or no opinion)

(If Yes)

5. Are the streets and alleys now:

_______ Somewhat cleaner _______ Somewhat dirtier

_______ Very much cleaner _______ Very much dirtier

Exhibit 10 (Continued)

(Ask #6 and #7 if a special cleaning or collection operation is being evaluated. Otherwise go to #8.)

6.* Do you know what Operation ___(Insert name of special operation)___ is?
_______ Yes
_______ No (_______ Not sure)

(If yes)
7.* Do you think Operation ___(Insert name of special operation)___ made a lasting effect on the cleanliness of streets and alleys:
_______ Yes
_______ No

8. Does the city of _______________________ collect your trash and garbage?
_______ Yes
_______ No (_______ Don't know. If no, or don't know, terminate unless private contractors are being evaluated.)

9. How often do the collectors miss picking up trash and garbage on schedule? Do they:
_______ Hardly ever miss (or never except for holidays and bad weather)
_______ Miss about once a month
_______ Miss more than once a month (_______ Don't know or no opinion.)

10. How often do the collectors spill trash and garbage or leave some that you have set out? Do they spill or leave some:
_______ Hardly ever (or never)
_______ About once a month
_______ More than once a month (_______ Don't know or no opinion)

11. How often do the collectors make so much noise that it bothers you?
_______ Hardly ever (or never)
_______ About once a month
_______ More than once a month (_______ Don't know or no opinion)

12. How often do you notice widespread odors from uncollected garbage?
_______ Hardly ever (or never) _______ Frequently
_______ Occasionally during hot weather (_______ Don't know or no opinion)

13. Have you been dissatisfied with your trash collection service at any time and for any reason in the last twelve months?
_______ Yes
_______ No

(If Yes, write in reason they gave: _______________________________________)

14. Have you complained to the city government about the cleanliness of streets and alleys or refuse collection services in the last twelve months?
_______ Yes
_______ No

15. How many times? _______________________

16. Were the conditions about which you complained corrected?
_______ Always
_______ Usually
_______ Seldom if ever

* For use if questionnaire is also to be used to help evaluate a special clean-up campaign.

The main points to look for in analyzing the data are significant changes in responses from the last survey and differences in responses among the areas throughout the city. Particular attention should be paid to how responses changed in areas that were subjected to special collection or clean-up programs. Once appropriate computer programs have been established, it should be possible for an analyst to collate and analyze the responses in less than one week.

There are a number of limitations to this type of survey which should be understood by those who use the findings.

- The bulk of the interviews are conducted by telephone from a sample whose responses may not be fully representative of the entire population. It does not include newcomers, transients, or persons with non-published numbers. The 50 in-person interviews give only a gross check of whether the responses of the telephone survey are representative. An estimate of the fraction of the total population in the non-represented categories should be made. If the fraction is small, the 50 in-person interviews should be sufficient. If the fraction is larger than perhaps 0.25, it may be desirable to have a larger number of in-person interviews.

- As only 100 persons will be interviewed in each service area, the survey cannot achieve the high degree of accuracy attained by national opinion polls that question hundreds of people. However, we do not believe urban administrators need such precision for making decisions.

- The telephone survey responses may not be highly accurate on citizen views, perhaps because the respondents do not have the time to think, their memories are poor, they are afraid to give a derogatory answer for fear of repercussions, or they give any answer just to get the survey over with.

- People may give intentionally biased answers to try to influence the actions of the sanitation department.

The willingness of D. C. citizens to participate seems to indicate, at least in this instance, that the last two factors were not especially relevant. If the limitations are recognized and the sampling and surveying procedures

Exhibit 11. ESTIMATED COSTS AND MANPOWER REQUIREMENTS FOR A MINIMUM-COST ANNUAL CITIZEN ATTITUDE SURVEY

Function and Personnel[a]	Number of Responses: 650		950		1450	
	Man Weeks	Cost	Man Weeks	Cost	Man Weeks	Cost
Preparing Telephone Calling Lists						
Clerk or Secretary	2	$ 400	3	$ 600	4½	$ 900
Telephone Interviewing (overtime work)						
2 Clerks or Secretaries	1 1/3	$ 400	2	$ 600	3	$ 900
Analysis and Preparation of Special Reports						
Analyst	1	$ 400	1½	$ 600	2½	$ 900
50 In-Person Surveys @ $25 per Interview		$1,250		$1,250		$1,250
Total	4 1/3	$2,450	6½	$3,050	10	$3,950

a. Personnel costs include employee salary and benefits and overhead. Clerk or secretary annual total costs were estimated

are performed carefully, the information should be sufficiently accurate for obtaining reasonable estimates of citizen perception and changes in perception over time.

MANPOWER REQUIREMENT, COSTS

The estimated manpower requirements and cost of a citizen attitude survey for several different size surveys are shown in Exhibit 11. Note that the low costs per interview (as compared to more typical field survey costs) are due to the very short interview time and use of the telephone survey.

V

USES OF THE MEASUREMENT SYSTEM

QUESTIONS THAT IT CAN HELP ANSWER

How clean is the city? Are some parts significantly dirtier than the rest? How effective have special corrective programs been? Are conditions changing over time? Is trash and garbage stored adequately and collected frequently enough? What proportion of the citizens are generally satisfied with refuse collection services? What aspects trouble them?

Such questions can be answered at least in part through analysis of appropriate data gathered with the measurement system that we have outlined and field tested.

We identify six major types of data needed by local government officials regarding the effectiveness of solid waste collection services and which may be obtained through the measurement system:

1. Periodic assessment of the overall effectiveness of solid waste collection services and identification of general problem areas.

2. Evaluation of special solid waste collection programs.

3. Allocation of collection resources throughout the community.

4. Analyses of alternative ways to provide solid waste collection services.

5. Monitoring performance of collection personnel and providing basic information for possible use in labor contracting.

6. Regulation of private solid waste collection contractors.

Each is discussed below. As noted earlier, it should be emphasized that the measurement system does not attempt to discover why conditions are as they are, why they have changed, or what should be done to correct problem conditions. Its purpose is to indicate what the conditions are (in various sections of the community as well as the whole), and how they change over time. The measurement information nevertheless should provide important clues which, along with other information, can indicate why a problem exists and what might be done to alleviate it.

Other specific uses for the measurement system are possible, particularly if the inspection data collection format is expanded. For example, inspectors might also be requested to obtain such information as the frequency of spillage or scattering of waste from collection operations or the overfilling of storage containers. These could be used to help determine whether the frequency of collection is adequate.

PERIODIC ASSESSMENT OF OVERALL EFFECTIVENESS OF SERVICES AND IDENTIFICATION OF PROBLEM AREAS

Periodic assessment is the central purpose of the measurement system. The formats presented in Exhibits 7, 8, and 9 illustrate the types of outputs possible. In addition to such summary tables and maps, interpretive reports based on careful analysis of the data and other pertinent information on the collection situation should be provided to management by the measurement system analyst.

This material should be provided to public officials at least annually and possibly semi-annually, to indicate trends over time and variations in conditions among service areas in the community. The evaluative reports should highlight the major problems and show evidence of progress, or the lack of it.

EVALUATION OF SPECIAL CLEAN-UP PROGRAM

Special waste collection programs can be evaluated systematically with data from the measurement system. Such programs include tests of new equipment, new collection methods or procedures, and community publicity programs to reduce improper waste storage and handling by citizens. These evaluations may require specially scheduled field inspections or citizen surveys so that the timing and coverage of the measurements are compatible with those of the programs being evaluated.

To illustrate: In August 1971, the D. C. government initiated "Operation Clean Sweep," a citywide special collection and cleaning program in which several hundred supplemental cleaning personnel devoted over two months to cleaning the city. They spent about one week in each of the District's nine service areas. All streets and alleys were cleaned. Residents were encouraged to set out refuse for special collections. One of the principal purposes of Operation Clean Sweep was to test the concept that once the city is made clean, it can be kept clean.

The measurement system was used to help determine the amount of change in litter conditions resulting from Operation Clean Sweep and to guide future actions (e.g., should a modified or identical Clean Sweep operation be conducted at a later date).

There were two major parts to the evaluation – inspections of the entire city using the photographic rating guide, and a telephone survey of about 100 residents in one service area.

1. Inspection Results.

Using a version of the procedures described in Chapter III, inspection operations were undertaken in each service area the week prior to and the week after the sweep to measure the short-term effects on cleanliness of streets and alleys. A third inspection operation was scheduled approximately two months afterward to see what longer-term effects Operation Clean Sweep had on the general level of cleanliness throughout the District of Columbia. For each area, the average street litter rating and the percent of streets in the sample with ratings of 2.5 or worse were calculated – for each of the three sets of inspections. The procedure for comparing the "immediately before" and the "immediately after" conditions is illustrated in Exhibit 12. This exhibit is illustrative. For instance, it combines the nine District of Columbia service areas into six. But it closely approximates the kinds of findings that resulted.

We cannot be certain that statistically significant changes in the average litter rating are attributable to the special cleanup operation. However, if no good alternative explanations can be uncovered, there is considerable reason for attributing the difference in ratings to the clean-up operation.

It is possible for streets and alleys to be more littered after a clean-up campaign. This can happen if residents think that the clean-up program will last a long time and put out refuse for special collection after the program ends. The refuse becomes scattered by dogs and even people to an extent that the streets and alleys might be dirtier than before.

Exhibit 12. ILLUSTRATIVE EXHIBIT SHOWING EFFECTS OF CLEAN-UP PROGRAM ON STREET LITTER RATINGS

Service Area	Number of Blocks in District Area	Average Litter Rating[a] for Sample		Change in Litter Rating (+ shows improvement)	Statistically Significant Change[b]	Percent of Streets Rating 2.5 or Worse		Change percentage Points (+ shows improvement)	Statistically Significant Changes[c]
		Before CleanSweep	After CleanSweep			Before C.S.	After C.S.		
1	800	2.43	2.11	+.32	Yes	15	7	+ 8	No
2	700	1.80	1.79	+.01	No	0	0	0	No
3	1,100	3.13	2.53	+.60	Yes	61	22	+ 39	Yes
4	1,300	2.10	2.32	- .22	No	10	25	- 15	Yes
5	700	1.58	1.44	+.14	No	0	0	0	No
6	1,400	2.71	2.76	- .05	No	31	38	- 7	No
Total City	6,000	2.38[d]	2.27[d]	+.11[d]		29[d]	25[d]	+ 4	

a. The actual comparison made in the District of Columbia was more complex.

b. The statistical significance was established when the absolute difference in average ratings was 0.25 or more.

c. The statistical significance was established when the criterion in Appendix B was met.

d. Average conditions for the entire city were estimated by weighting the average litter rating (or the percent of streets rated 2.5 or worse) by the number of blocks in the service area, adding the weighted sum, and dividing by the total number of blocks in the city. Each service area was actually divided into census tracts and appropriate samples drawn in each tract. Thus, the Service Areas figures are weighted averages of each's sub-areas.

2. Citizen Survey Results.

A survey of 116 citizens in a selected portion of the District was conducted three months before Clean Sweep, and a similar survey was conducted about three months after Clean Sweep. The basic survey instrument used was a version of that shown in Exhibit 10. The follow-up survey was to measure changes in citizen satisfaction and perception of problems which could be attributed to Clean Sweep. It also contained some additional questions to measure citizen awareness of, and opinion of the effectiveness of Clean Sweep.

Exhibit 13 illustrates the type of information that is provided by the citizen survey for evaluating the Clean Sweep program. It presents illustrative, not actual, responses from only a portion of the questions contained in the full survey (see Exhibit 10). In the District of Columbia, the only area in which there was a statistically significant improvement in citizen opinion was in alley cleanliness. Questions 4-7 illustrate the type of queries added to help evaluate a one-time clean-up operation.

In general, for each evaluation an interpretative report based on a careful analysis of the measurement system data, along with other pertinent information on the circumstances of the special program, should be provided to management by the measurement system analyst.

GEOGRAPHIC ALLOCATION OF COLLECTION FORCES

Useful information can be provided by the measurement system for the calculation of force requirements for one-time clean-ups or for more prolonged operations.

The force requirements for a one-time cleaning of entire neighborhoods can be estimated readily from a map that shows litter conditions on a block-by-block basis if productivity data of cleaning forces is available. Exhibit 14 and the text accompanying it illustrate this use in the District of Columbia.

Average litter ratings for each service area, whether measured in a sampling inspection or a complete inspection, will document those areas of the community that have the greatest need for collection services. Local government officials can use this information along with other relevant factors in making their allocation decisions. For example, manpower

Exhibit 13. ILLUSTRATION OF INFORMATION OBTAINED FROM A CITIZEN SURVEY

Question:	Before "CleanSweep" Respondents (%)	After "CleanSweep" Respondents (%)	Significant Difference
1. What do you think of the cleanliness of the street you live on?			
a. Clean	14	13	
b. Mostly Clean	45	50	
c. Fairly Dirty	27	24	
d. Very Dirty	14	13	
e. No Opinion	0	0	
%Satisfied[a] (a and b)	59 ± 8	63 ± 7	No
%Dissatisfied(c and d)	41 ∓ 8	27 ∓ 7	
2. What do you think of the cleanliness of the alleys in your neighborhood?			
a. Clean	8	9	
b. Mostly Clean	24	30	
c. Fairly Dirty	21	26	
d. Very Dirty	32	26	
e. No Opinion	15	9	
%Satisfied (a and b)	34 ± 9	44 ± 8	Yes
%Dissatisfied(c and d)	66 ∓ 9	56 ∓ 8	
3. What do you think of the overall cleanliness of your neighborhood?			
a. Clean	14	9	
b. Mostly Clean	48	57	
c. Fairly Dirty	23	30	
d. Very Dirty	12	4	
e. No Opinion	3	0	
%Satisfied (a and b)	64 ± 8	64 ± 7	No
%Dissatisfied(c and d)	36 ∓ 8	36 ∓ 7	

---

a. The 90 percent confidence limits given with the percentages "satisfied" and "dissatisfied" were calculated using the formula

$$1.7\sqrt{\frac{P\ (I\ -\ P)}{n}}$$

where P = proportion of sample satisfied
n = number responding to the questions

Exhibit 13 (Continued)

Questions to Evaluate the Effects of a "Clean Sweep" Operation:		Responses (Percentage)
4. Have you noticed any change in the cleanliness of streets and alleys in your neighborhood in the last 3 - 4 months?		
Yes		43 ± 8
No		52 ∓ 8
Don't Know/No Opinion		5
5. (If Yes) Are the streets and alleys now:		
a. Somewhat Cleaner		68
b. Very Much Cleaner		22
c. Somewhat Dirtier		6
d. Very Much Dirtier		4
TOTAL		100
Satisfied	(a and b)	90 ± 7
Dissatisfied	(c and d)	10 ∓ 7
TOTAL		100
6. Do you know what Operation Clean Sweep was?		
Yes		67 ± 7
No		30 ∓ 7
Not Sure		3
TOTAL		100
7. Do you think Operation Clean Sweep made a lasting effect on the cleanliness of streets and alleys?		
Yes		43 ± 8
No		57 ∓ 8
TOTAL		100

could be assigned in proportion to the litter rating. This might involve taking men from relatively clean areas. The actual rating values can be used to help justify the reduction in forces. The effects on cleanliness of different levels of force allocation can be made by analyzing changes in average litter ratings over time as a function of forces allocated. This information would be especially useful in allocating forces in the future.

COMPARING ALTERNATIVE WAYS TO PROVIDE SERVICES

The real long-term payoff for the measurement system may lie in helping to evaluate different ways to improve the existing system. The measurement system will provide data immediately useful in providing a baseline of service effectiveness against which alternative ways should be compared. Data collected over time should be useful in identifying causal relationships between collection programs or activities and their effects on the quality of service. Thus, it should be possible to extrapolate results from pilot tests of new equipment, new procedures, new community publicity programs, etc. When this data is analyzed along with cost information, it should be possible to estimate more reliably what would be the costs and effects if such new equipment, new procedures, etc., were more widely instituted in the community.

Unfortunately, the scope of our work has not permitted development of examples of this important use of the measurement system data.

ASSESSING COLLECTION PERSONNEL PERFORMANCE

Although this use was not tested in the District of Columbia, it is possible to monitor performance and to provide feedback to specific collection crews and foremen on the results of the inspections and citizen surveys for their respective geographic areas. (To the extent that service areas may be covered by more than one crew, such feedback would have less relevance.) It is rare today that public or private employees can learn the impact of their work on the community, as could be the case here. Such feedback might well help to motivate all concerned to improve their performance. Of course, major factors other than collection crew performance affect actual and perceived cleanliness – such as citizen behavior, equipment, and working conditions.

Another untested possibility of the inspection data is to use it along with workload data to develop performance standards and incentive plans for solid waste collection workers. These would need to be worked out mutually with the appropriate unions. At this point we are necessarily vague on how this might be done, but we point to it as a possible use when and if a city and its employees develop confidence in the measurement outputs.

SETTING PERFORMANCE STANDARDS TO REGULATE PRIVATE SOLID WASTE COLLECTION CONTRACTORS

The measurement system's data would be useful to local governments that contract directly with private contractors for all or part of their solid waste collection services, or which have the authority to regulate such contractors. Given the data, the local government could monitor the overall performance and effectiveness of private contractors to ensure the meeting of minimum quality levels. Performance standards relating to the measures in the measurement system might even be included in the contractual arrangements. In this fashion the local government would be able to obtain objective and representative information on the degree to which the private contractor was servicing the community. This use has not been tested. Again, the details would have to be worked out, probably after community confidence in the quality of the measurement system data has been created.

Exhibit 14. STREET LITTER CONDITION CLASSES AND SQUARE MILES IN EACH[a]

(Map on facing page: Portion of Washington, D. C., area east of Anacostia River)

Litter Condition Class[b]	Square Miles Per Class[c]	Percent of Total Area
☐ 1.0-1.5[d]	3.80	39
☐ 1.6-2.0[d]	3.23	34
2.1-2.5	1.98	21
2.6-3.0	.50	5
3.1-3.5	.10	1
Total	9.61	100

Use of Map: A map of this sort, in conjunction with other information, can be used to estimate manpower requirements for street cleaning operation. For example, assume that city officials decide that the areas of worst conditions (the two most darkly shaded areas, with litter ratings from 2.6 through 3.5) will be upgraded to the ratings of the two best classes (1.0-1.5 and 1.6-2.0). Total area involved is .6 square miles, and total curb length of streets (both sides) is found to be 187,000 linear feet. Using productivity data of the D. C. Department of Environmental Services for cleaning streets, sidewalks, and tree box areas, it is calculated that the goal would require four crew days and two mechanical sweeper days, and an additional two flusher crew work days for flushing gutters.

a. Size and scale permits display here of only one-third of the actual map devised for this analysis. The data and discussion below them relate to the larger area—all of the District of Columbia east of the Anacosta River.

b. Condition classes are based on block litter indexes reported April 26-28, 1971. These indexes are the averages of all street inspection ratings for each block. (Alleys not considered in this map and analysis.)

c. The total area and portions in each class measure all sectors where the District of Columbia government has street cleaning responsibility but exclude sectors under jurisdiction of the National Park Service, Bolling Air Force Base, and so forth. See also Note a.

d. The two cleanest conditions are not differentiated on this map in order to show more clearly the areas that are moderately and heavily littered.

HOSPITAL
ELIZABETHS
NICHOLS
ADAS ISRAEL HEBREW CEMETERY
CAFRITZ HOSPITAL
LEBAUM
STREET
WILSON
MELLON
NEWCOMB
OAKWOOD
ORANGE
PORTLAND
ALABAMA
WHEELER
CONGRESS
MISSISSIPPI
Oxon Run
VARNEY
RALEIGH
BROTHERS
HIGHVIEW PL.
SAVANNAH
BALLOU HIGH SCHOOL
HART JUNIOR HIGH SCHOOL
OXON RUN RECREATION CENTER
BALTIMORE & OHIO R.R. CO.
OVERLOOK
ANACOSTIA FREEWAY
WAYNE PL.
XENIA ST
YUMA
ATLANTIC
LIVINGSTON
CONDON
BARNABY
BRANDYWINE ST.
SOUTH CAPITOL
CHESAPEAKE
DANBURY
ELMIRA
GALVESTON
CAPITOL
FORCE BASE
PARKWAY
Parkway
HADLEY HOSPITAL
NAVAL HOUSING PROJECT
MAGAZINE
ROAD
AVENUE
JOLIET
Bald Eagle Hill

VI

SOME MISCELLANEOUS CONSIDERATIONS

Three other important considerations in a decision to implement a solid waste collection effectiveness measurement system are:

- Keeping costs in both dollars and manpower to a minimum.
- Dissemination of information.
- Organizational location of the system responsibility.

KEEPING COSTS TO A MINIMUM

1. Use of Existing Personnel for Inspection.

For the most typical versions of the system, two different inspectors will be required, each working intensely during certain parts of the year – probably at most one-third or one-half of a full year for each man. In many local governments it should be possible to utilize existing personnel for these inspector positions, thereby avoiding additional expenditures. Certainly this possibility should be explored.

2. Use of Volunteers for Inspection.

If existing personnel or funds to hire new personnel are not available, a second possibility is to use citizen volunteers for the inspection process (not tested in the D. C. experiment). The part-time nature of the inspection task makes this feasible, although the need to make inspections during daylight hours will probably reduce the availability of volunteers. However, weekend inspections are possible, and would be appropriate for many objectives. In either case, as discussed in Chapter III, the training and replication of the inspectors is essential to assure that they are able to follow the photographic rating guide closely enough.

3. Obtaining a Qualified Analyst.

The most difficult resource constraint may well be the analyst. He is responsible for establishing the inspection and citizen survey sampling procedures, for assuring that the inspector ratings are kept in conformance with the photographic rating guide, and for analyzing the resulting data. Most local governments today lack experienced analytical capability. The demands for this system are not very great, however. We believe that a recent college graduate, perhaps with a Masters degree in public administration, who has had at least a minimum quantiative background (particularly one or two courses in statistics) should be able to handle the major aspects of the system. For certain specialized sampling matters, he might need to obtain some assistance from others in the government, a nearby university, a private firm, or citizen volunteers.

4. Use of Internal or Volunteer Help for Telephone Interviewing.

For the telephone citizen survey interviewing, the time required as described in Chapter IV would be about five to seven man-weeks of effort for each annual survey of about 600 to 900 families. The measurement system analyst might want to undertake some of these interviews. Other government personnel, such as clerks and secretaries, should be able to conduct the interviewing without a great deal of training. If for some reason city or county employees are not available, it may be possible to use citizen volunteers. Again, some training and checking of the telephoning should be undertaken by the analyst. Our past experience indicates that this should not be a major problem.

5. Use of Cluster Sampling to Reduce Inspection and Travel Costs.

Another possible cost reducing approach is related to the sampling procedures for the inspections. If block-faces to be sampled are selected at random from all blocks in a service area, the travel time during inspection will be relatively large, thus increasing costs. Many major public opinion polling firms are using a "cluster-sampling" approach to reduce such costs. They randomly select a number of locations in a jurisdiction and for each of those locations, interview a number of households (or, in our case, inspect block-faces) within the immediate vicinity of that location. This was the approach actually used in our work in the District of Columbia. For each location, four to seven geographically consecutive block-faces were inspected. The drawback is that for a fixed number of block-faces, if some are clustered, the precision of the result will be less than if each individual block-face is randomly selected. Thus, more block-faces would need to be sampled in cluster sampling to achieve the same degree of precision. The technical aspects of this choice and the tradeoffs in cost and precision are beyond the scope of this paper. If a jurisdiction does find that travel time and costs seem to be excessive with random sampling, it should consider cluster sampling as a possibility. Expertise should then be sought to identify the options to the government.

DISSEMINATION OF MEASUREMENT SYSTEM FINDINGS

Information on the cleanliness of streets, obtained through the inspection procedures and the citizen surveys, need to be released to the public. Clearly, governments can expect reaction to these results, particularly from neighborhoods that are dissatisfied with the findings. This is likely to be a problem particularly when the initial inspection and survey are undertaken. However, in the long run we suspect that it will be to everyone's benefit and may indeed permit the government to obtain community participation to help improve cleanliness conditions.

One caution we do suggest. Generally, inspections and surveys should not be scheduled just prior to an election. They would then have special impact, and officials could be accused of using the inspections and surveys for primarily political purposes.

ORGANIZATIONAL CONSIDERATIONS

An important question is where to locate the measurement system organizationally in the local government. As a general principle, it is not advisable to place responsibility for the evaluation of programs with the persons directing the programs that are being evaluated. On the other hand, the measurement system should be administered as closely as possible to the department in charge of solid waste collection.

The type of organizational arrangement currently existing in the District of Columbia seems an appropriate compromise. The measurement system is currently the responsibility of the Department of Environmental Services and essentially placed in a special planning and analysis office in that department. The head of the solid waste collection operation is a key member of the advisory group for the measurement system, and his solid waste collection unit is the major user of the data from the system.

In addition to providing information for collection planning and operations, there should be explicit provision for the inspection survey results to be made available rapidly to other city officials, including the mayor or chief executive and local legislative officials.

Appendix A

DETAILED INSTRUCTIONS FOR INSPECTORS

The ratings of streets and alleys, and if so desired, private property, should be conducted by trained inspection personnel. Their inspections should be replicated periodically by the analyst in charge of the program to assure that each rating value always represents the same degree of cleanliness. The analyst should provide the inspectors with maps showing the locations of the block-faces and alleys that are to be inspected, and showing the route or routes that the inspectors must follow to make these inspections. Basic information that is to be collected for each inspection was shown in Exhibit 3, and the procedure for tape recording this information is given in Exhibit 15. We have assumed that the information will be dictated into a tape recorder and that the inspector will have been given appropriate instructions for using the tape recorder. Data could be collected and recorded on a written form as illustrated in Exhibit 6. But this would slow down the inspection procedure somewhat because the inspector would be required to stop at the end of each block-face or alley to record the data.

In the field an inspector inspects block-faces or alleys following a route specified by the analyst. He drives to the beginning of the route, which should be drawn on a map of the area. After checking his recorder to make sure it is working, the inspector records his identification number, date, time, area of the inspection, and route number on the tape. The inspector drives along the route indicated on the map, making his inspection report. The report for each block-face is prefaced by the block-face identification number. The part of the public way being reported for a given block-face is from the middle of the street to the right across the curb and sidewalk area to the property line, and is bounded on either end by the cross streets.

Using the four-point rating scale with half-point intermediate values, he reports the average conditions on the block-face. When more than half of the litter is leaves, the word "leaves" will be reported along with an estimate of what the average litter condition of that block-face would be without leaves. When the average condition over the entire block-face is 1.0, 1.5 or 2.0, but one or more segments – each 100 feet long – have conditions of 2.5 or worse, the worst condition observed would be reported along with the average. For example, if the worst 100-

Exhibit 15. INSPECTION REPORTING PROCEDURE[a]

At beginning of each cassette side:

Description	Examples of words recorded
Inspector number	Inspector one
Date of inspection	August 9, 1971
Weather: Indicate whether clear, windy, rain, snow. Indicate change at beginning of tract or traverse. For all inspector observations to follow on a given cassette side	Weather clear and windy
Time inspection begins	Time: 9:20
Service area number	Service area Six
Indicate whether streets or alleys are being inspected	Streets
Tape cue	Double cue
Census tract number	Beginning tract 48.2
Location (street, intersection or junction)	7th and N Streets
Tape cue	Single cue
Traverse number	Traverse one
Block-face identification number	Block-face one seven five
Average condition	Condition two point five
Presence of leaves	Leaves
Estimated average condition without leaves	Estimate condition two

a. This is the procedure used in the District of Columbia. Some of the terms developed for that study might be used elsewhere. For example, a "tape cue" is an electronic "beep" signal used with the particular recorder employed; a "segment" is defined in Appendix A; a "traverse" is a series of continuous block-faces on the same side of a given route; and codes were used to designate special items or conditions, e.g., Code 1 – Abandoned auto; Code 2 – dead animal, and so forth.

Exhibit 15 (Continued)

If average condition is 1, 1.5 or 2 and there is a Condition 2.5 thru 4 segment	Segment three on public way
On public way, report that condition Report only the worst condition	
If a segment with a worse condition is observed later on the same block-face make the correction	Correction-Segment four on public way
If average condition is 3, report presence of 3.5 or 4 segment on public way	
If average condition is 3.5 report presence of 4 segment on public way	
Report presence of condition 2.5 thru 4.0 Segment area beyond public way (whichever is worse). Report only the worst one, no matter how many there are. We are not trying to count the number of bad spots-only indicate that there is one or more bad spots.	Segment four beyond public way
Coded items or conditions (See code sheet for numbers and definitions)	Code 1, Code 2, Code 9
Second block-face in traverse	Block-face one seven two Condition three point five Segment four on public way Code six
Third block-face in traverse	Block-face one seven three Condition two
Continue reporting block-face information. The report for the last block-face in the traverse would be like this:	Block-face one seven four Condition two point five Segment three beyond public way

Exhibit 15 (Continued)

Code four
Code three
End of traverse one
Single cue

The process continues through all traverses (and block-faces) in the tract. At the end of the last block-face reported for that tract the inspector says:

End of traverse Six
End of tract 48.2
Time 10:22
Double cue

At the beginning of next tract the inspector announces new tract saying:

Beginning tract 48.1
Time 10:30
5th and O Streets
Single cue
Traverse one
etc. as before

In case segmenting of a block-face is required the inspector would report saying:

Block-face six seven three
Segmenting: two
two point five
one
one
four
End of Segmenting
Three segment beyond public way

foot segment happened to be "3.5" on a block-face with an overall rating of "2.0," the statement dictated would be "segment 3.5 public way, average rating 2.0."

In addition to the litter rating data, the inspector also reports specific items or conditions on each block face using the code identified in the detailed instruction sheets. Variations in reporting the average litter condition seem likely to occur in those cases where blocks are long or the conditions within a block are so varied that the inspector is unable to make a statement about the average condition without going through a segmenting procedure. It consists of reporting the condition of a block-face or alley by a series of segment ratings describing the litter conditions for each 100-foot linear segment of a block-face or alley. In such a case, the inspector does not report the average condition but rather prefaces his inspection ratings with the work "segmenting."

Using the four-point scale with intermediate ratings, he proceeds to record a rating of 1.0 through 4.0, whichever is appropriate for each 100-foot segment. He then reports termination of segmenting for the block-face by saying "end of segmenting." In such a case, no average litter rating would be reported. The average rating is calculated by the clerk transcribing the tape.

In recording alleys, the details reported would be the same as for block-faces except for the labeling for the alleys and the use of alley numbers. Under normal circumstances, street inspection operations should be separate from alley operations. Such specialization of activity simplifies operational planning and control, makes most efficient use of total inspection time, and also simplifies reporting, transcription, data processing, and analysis.

Appendix B

SAMPLE SIZE DETERMINATION AND SIGNIFICANCE TEST

We propose that the sample size be sufficiently large to detect changes of 0.25 or more in the average litter rating at a 90 percent confidence level. In other words, the sample size should be the number of block-faces or alleys such that, if the average litter ratings of two equal size samples (either from two different areas or the same area but at different times) are compared and the absolute difference is 0.25 or more, one can conclude that in 90 cases out of 100 the difference is due to changes in the average litter rating of all the block-faces in the area and not due to the chance selection of certain block-faces.

Formula A in Exhibit 16 can be used to determine the sample size for this criterion, whether it be for alleys or for block-faces. The sample size depends upon the number of block-faces (or alleys) in the area and upon the estimated variance – a statistical measure of the uniformity of the values of the ratings. The Table in Exhibit 16 shows sample size as a function of the total number of blocks in the area and of the estimated variance.

The actual value of the variance will not be known until after the sample has been taken. However, it is suggested that the variance be assumed to be 0.30 and that 30 block-faces or alleys, as the case may be, be selected in a statistically random manner for each area sampled. Then the variance should be estimated using Formula B. If the estimated variance is greater than 0.30, Formula A or the Table should be used to determine how many block-faces or alleys must be inspected for that variance.

Exhibit 16. THREE FORMULAS FOR CHOOSING SAMPLE AND ANALYZING DATA

FORMULA A

N is the sample size to detect a 0.25 or greater difference in average litter ratings of two areas or the same area at different times at 90 percent confidence level.

$$N = \frac{1}{\frac{1}{N_o} + \frac{1}{95 \text{ var } x}}$$

where

N_o = Total number of block-faces (or alleys) in the area being sample-inspected

var x = Estimate of the variance of the ratings of the block-faces (or alleys) in the area

Table on Selected Values of N

Total Number of Block-Faces (or Alleys) in the Area Being Sampled	Variance of Litter Ratings			
N_o	.15	.30	.50	.75
100	13	22	32	42
250	14	26	40	56
500	14	27	44	62
1000	14	28	46	67
>>1000	14	29	48	71

FORMULA B

Estimate of the variance of the block-faces (or alleys) ratings in the area from inspecting only N block-faces (or alleys).

$$\text{var } x = \frac{\sum^{N} (x - \bar{x})^2}{N - 1}$$

where x = Actual rating of a block-face (or alley)

$\bar{x}$ = Average rating of all block-faces (or alleys) with sample

N = Number of block-faces (or alleys) inspected

Exhibit 16 (Continued)

The additional number should be selected at random and inspected. Then the average litter ratings from the sample size should be found and used as the estimate of the average litter rating for the entire area.

Often it will be desirable to know if there has been a statistically significant change in a percentage rating. For example, was a shift in the percent of citizen responses to a question in the survey significant or was the decrease in the percent of streets with ratings of 2.5 significant. Statistically significant[a] shifts can be detected when the inequality in Formula C is met. By illustration the terms of the formula are for street litter ratings in excess of 2.5. The P's and N's could just as well have been described for citizen response.

Formula C

Significance test.

$$\frac{|P_1 - P_2|}{100} \geq 1.7 \sqrt{\frac{P_1\left(1 - \frac{P_1}{100}\right)}{100 \quad N_1} + \frac{P_2\left(1 - \frac{P_2}{100}\right)}{100 \quad N_2}}$$

Where

P_1 = Percent of streets inspected before the cleanup that rated 2.5 or worse

N_1 = Number of streets inspected before cleanup

P_2 = Percent of streets inspected after cleanup that rated 2.5 or worse

N_2 = Number of streets inspected after cleanup

a. At the 90 percent confidence limit.